DINOSAURS

BRACHIOSAURUS

BY PRISCILLA AN

Kids Core
An Imprint of Abdo Publishing
abdobooks.com

abdobooks.com

Published by Abdo Publishing, a division of ABDO, PO Box 398166, Minneapolis, Minnesota 55439. Copyright © 2024 by Abdo Consulting Group, Inc. International copyrights reserved in all countries. No part of this book may be reproduced in any form without written permission from the publisher. Kids Core™ is a trademark and logo of Abdo Publishing.

Printed in the United States of America, North Mankato, Minnesota.
102023
012024

THIS BOOK CONTAINS RECYCLED MATERIALS

Cover Photos: Shutterstock Images (dinosaur, background)
Interior Photos: Elena Duvernay/Stocktrek Images/Science Source, 4–5; Richard Bizley/Science Source, 6; Mark Garlick/Science Source, 9, 26; Shutterstock Images, 10, 17 (left), 17 (middle), 28–29; Phil Wilson/Stocktrek Images/Science Source, 12–13; Thomas Banneyer/dpa/picture alliance/Getty Images, 15; Rimma Rii/Shutterstock Images, 17 (right); Akkharat Jarusilawong/Shutterstock Images, 18; Eric Nathan/Alamy, 20–21; Elmer Riggs/Field Museum Library/Premium Archive/Getty Images, 23; Anadolu Agency/Getty Images, 24

Editor: Marley Richmond
Series Designer: Mary Shaw

Library of Congress Control Number: 2023939657

Publisher's Cataloging-in-Publication Data

Names: An, Priscilla, author.
Title: Brachiosaurus / by Priscilla An
Description: Minneapolis, Minnesota: Abdo Publishing, 2024 | Series: Dinosaurs | Includes online resources and index.
Identifiers: ISBN 9781098292669 (lib. bdg.) | ISBN 9798384910602 (ebook)
Subjects: LCSH: Dinosaurs--Juvenile literature. | Prehistoric animals--Juvenile literature. | Brachiosaurus--Juvenile literature.
Classification: DDC 567.90--dc23

CONTENTS

CHAPTER 1
The Long-Necked Dinosaur 4

CHAPTER 2
Like a Giraffe 12

CHAPTER 3
Brachiosaurus Fossils 20

Dino Details 28
Glossary 30
Online Resources 31
Learn More 31
Index 32
About the Author 32

Brachiosaurus is just one kind of long-necked dinosaur. Long-necked dinosaurs are called sauropods.

THE LONG-NECKED DINOSAUR

The ground trembles. Small rocks bounce off the land. Tiny creatures skitter away. They avoid a herd of *Brachiosaurus* (BRAK-ee-uh-SOHR-uhs) walking nearby. Each dinosaur is about 40 feet (12 m) long.

Brachiosaurus ate many kinds of trees.

Their feet are massive. Their tall necks stretch high into the clear blue sky.

A couple *Brachiosaurus* glance down at the fleeing animals, but the *Brachiosaurus* aren't interested in eating them. This is because *Brachiosaurus* are herbivores. They eat only plants. Instead, the dinosaurs keep walking toward a forest of pine trees.

When the *Brachiosaurus* herd reaches the forest, the dinosaurs immediately start **grazing**. One *Brachiosaurus* approaches a tree. The dinosaur can reach the very top. It stretches its neck and tears off the pine needles. It swallows them whole.

Age of the Dinosaurs

Brachiosaurus is just one of many known dinosaurs. **Paleontologists** say that the first dinosaurs lived 245 million years ago. From then, dinosaurs lived on Earth for about 180 million years until they became **extinct**. *Brachiosaurus* lived about 150 million years ago.

The Extinction of Dinosaurs

Many scientists believe that the extinction of dinosaurs happened 66 million years ago. They think it was caused by an asteroid. Asteroids are huge rocky objects in space. The asteroid crashed into the coast of what is now Mexico. It affected many living things. One of its effects was causing huge waves to wash over the land.

The asteroid that may have caused dinosaurs to become extinct was 6 to 9 miles (10–15 km) wide. The crater it left behind was about 93 miles (150 km) across.

Scientists think *Brachiosaurus* would have held its long neck mostly upright while walking.

Dinosaurs varied in size and shape. Some were as big as *Argentinosaurus*, a dinosaur that was 120 feet (36 m) long and weighed about 100 tons (90 metric tons). Some were fierce meat eaters, such as *Tyrannosaurus rex*. Some walked on four legs and ate plants, such as *Brachiosaurus*. Scientists continue to learn more about dinosaurs today.

Explore Online

Visit the website below. Does it give any new information about dinosaurs that wasn't in Chapter One?

What Is a Dinosaur?

abdocorelibrary.com/brachiosaurus

Many other dinosaurs lived at the same time as *Brachiosaurus*.

CHAPTER **2**

LIKE A GIRAFFE

Brachiosaurus is often compared to a giraffe. Like giraffes, this dinosaur had a long neck for grazing treetops. It was an herbivore. It also had longer forelegs than hind legs. It stood more upright than many other four-legged animals.

This is how the dinosaur got its name. *Brachiosaurus* means "arm lizard." It refers to the dinosaur's long front legs.

Brachiosaurus lived in the area that is now North America. This dinosaur spent most of its time on flat land. That is because *Brachiosaurus* could weigh up to 62 tons (56 metric tons).

Water Dinosaurs?

Many scientists used to believe that *Brachiosaurus* lived in the water. The dinosaur's nostrils were at the top of its head. This could have made it easy for *Brachiosaurus* to breathe while walking on the bottoms of lakes and rivers. However, this belief was later proven wrong. *Brachiosaurus* had large pockets of air inside its body. Swimming or standing in deep water would have been difficult.

Having nostrils on top of its head may have helped *Brachiosaurus* breathe while eating and drinking.

This is the weight of about eight large elephants. *Brachiosaurus* was too heavy to walk up steep hills.

Despite its large size, *Brachiosaurus* had a small head. Paleontologists have found that *Brachiosaurus* had a brain the size of a tennis ball. Although *Brachiosaurus* was not the smartest dinosaur, it didn't have any **predators**. *Brachiosaurus* was too big to be hunted.

Plants and Teeth

Scientists are still learning about *Brachiosaurus*. Many think that these dinosaurs fed on tall **conifer** trees. Some think they fed on a plant called *Equisetum*. This plant was smaller, typically 2 to 5 feet (0.6–1.5 m) tall.

As Tall as the Trees

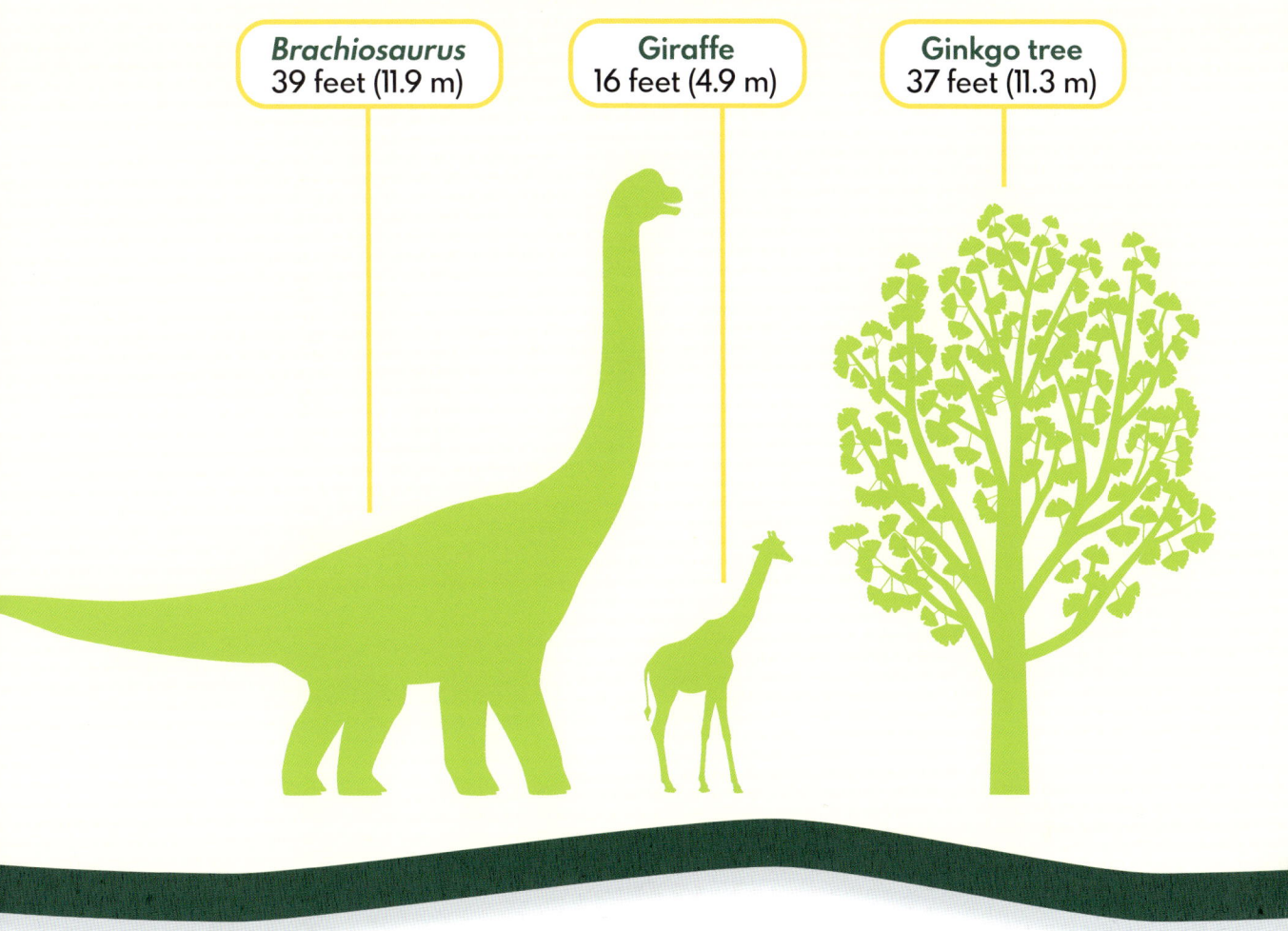

Brachiosaurus was about as tall as its food sources, such as an average-size ginkgo tree. These dinosaurs would have been two and a half times taller than the average giraffe today.

Brachiosaurus had simple, rounded teeth. They were not sharp. Scientists think this is because *Brachiosaurus* didn't chew its food.

Dinosaur teeth can show scientists important things about a dinosaur's diet and how it ate its food.

Instead, its teeth were used to tear leaves and bark from plants. Then *Brachiosaurus* would gulp its food down like a vacuum.

PRIMARY SOURCE

David Wilkinson is a scientist. He said *Brachiosaurus* was so large that walking uphill would use a lot of energy. It would climb hills only for food. He said:

> They're not going to climb hills unless by climbing it they are accessing a really good source of energy.

Source: Joseph Castro, "Brachiosaurus." *LiveScience*, 16 Mar. 2016, livescience.com. Accessed 7 Apr. 2023.

Comparing Texts

Think about the quote. Does it support the information in this chapter? Or does it give a different perspective? Explain how in two to three sentences.

The Field Museum of Natural History in Chicago, Illinois, has a life-size *Brachiosaurus* sculpture. The skeleton is modeled on the bones found by Elmer Riggs.

CHAPTER 3

BRACHIOSAURUS FOSSILS

The first *Brachiosaurus* **fossil** was found in 1900 by Elmer Riggs. Riggs was a paleontologist. He found a partial skeleton of the dinosaur in Grand River Valley in Colorado. He was the one who gave *Brachiosaurus* its name.

Since then, more *Brachiosaurus* fossils have been found across North America, in states including Utah, Oklahoma, Wyoming, and Colorado.

In May 2019, a *Brachiosaurus* fossil was found in Utah. The fossil was a bone from a foreleg. The bone was more than 6.5 feet (1.9 m) long. It weighed almost 1,000 pounds (453 kg). Because the fossil was found in a hard-to-access location, it wasn't moved by car. Instead, the bone was pulled out by two horses.

Finding a *Brachiosaurus* fossil is very rare. Only ten have been found. None of them create a complete *Brachiosaurus* skeleton. These fossils are displayed in museums such as the Utah Field House of Natural History State Park Museum.

Paleontologists from the Field Museum helped Elmer Riggs. A paleontologist compares himself to the *Brachiosaurus* foreleg bone the team discovered.

Brachiosaurus in Film

Although few *Brachiosaurus* fossils have been found, *Brachiosaurus* is a popular dinosaur.

A model *Brachiosaurus* was used to film *Jurassic Park III*. It was displayed at the Jurassic World Film Exhibition in China.

This is because it was featured in the movie *Jurassic Park*. The character Alan Grant is a paleontologist in the movie. The first time he sees a *Brachiosaurus,* Grant stands in awe of the humongous dinosaur.

However, there were a couple of differences between the movie's *Brachiosaurus* and the actual dinosaur. The movie shows the dinosaur rearing up to stand on its hind legs. In reality, a *Brachiosaurus* could not have done this, even for a very short amount of time.

Brachiosaurus or Giraffatitan?

In 1914, a paleontologist named Werner Janensch thought he found a *Brachiosaurus* fossil in modern Tanzania. But he was wrong. Janensch found a fossil of *Giraffatitan*. For many years, paleontologists thought *Giraffatitan* and *Brachiosaurus* were the same species. It was only in 2009 that scientists classified *Giraffatitan* as a separate dinosaur.

Scientists continue to update their ideas about dinosaurs' appearances, behaviors, and diets as they learn more about these ancient creatures.

Jurassic Park also shows the wrong dinosaur. When the movie was made, scientists believed *Brachiosaurus* and *Giraffatitan* were the same dinosaur. But scientists later discovered they were different **species**. The movie based its *Brachiosaurus* on *Giraffatitan*. This dinosaur was narrower than *Brachiosaurus*. *Giraffatitan* also had a shorter torso and tail.

People continue to learn new things about dinosaurs. About 45 new dinosaur species are discovered every year. Fossils will continue to show scientists more about dinosaurs, including *Brachiosaurus*.

Further Evidence

Look at the website below. Does it give any new evidence to support Chapter Three?

Where Can You Find Fossils?

abdocorelibrary.com/brachiosaurus

DINO DETAILS

Short tail used to balance the weight of the dinosaur's front end

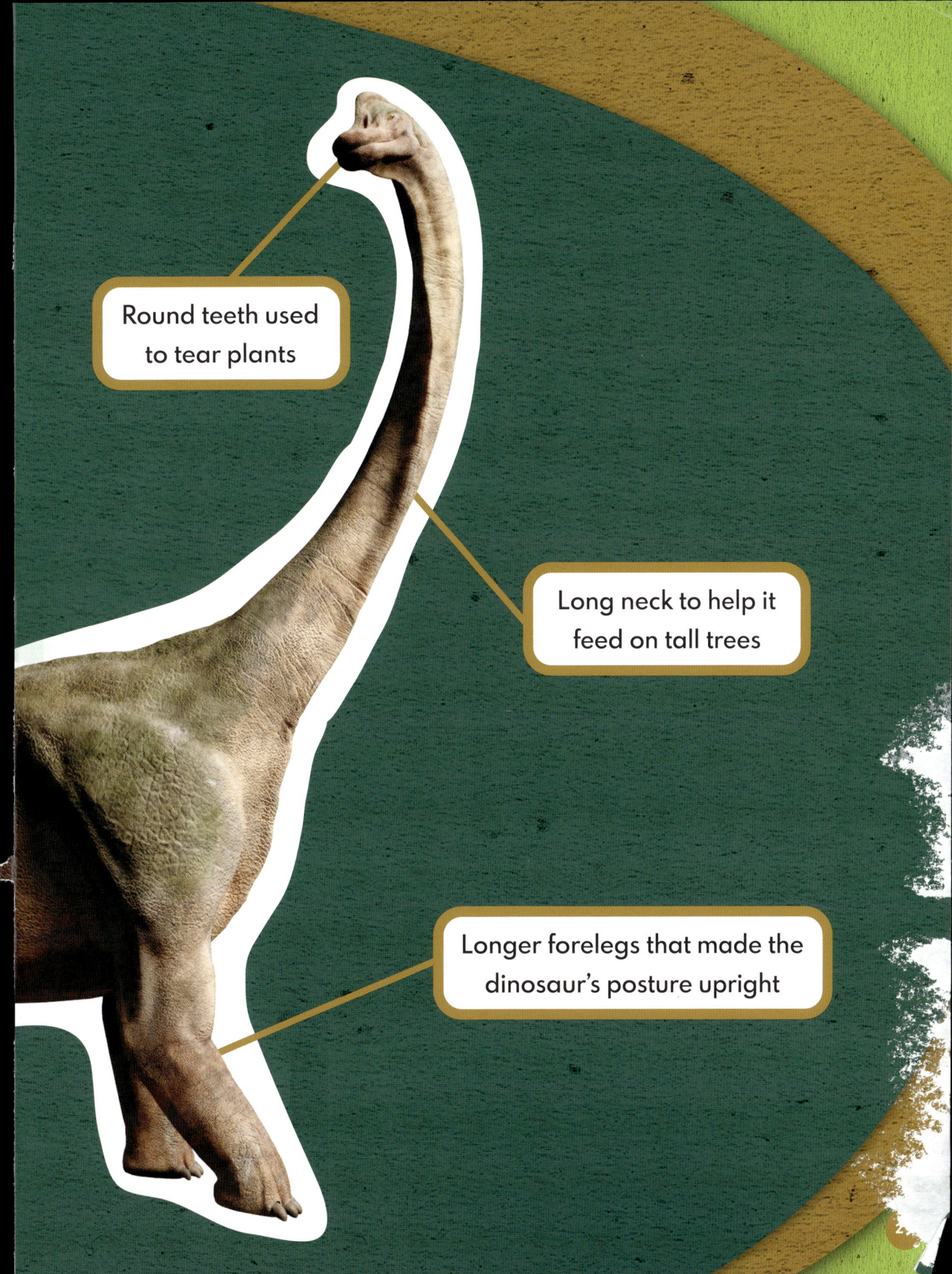

Glossary

conifer
a type of tree that has cones and needlelike leaves

extinct
no longer exists

fossil
the remains of very old animals or plants

grazing
feeding on plants

paleontologist
a scientist who studies fossils

predators
animals that hunt other animals

species
a group of the same kind of animals or plants

Online Resources

To learn more about *Brachiosaurus*, visit our free resource websites below.

Visit **abdocorelibrary.com** or scan this QR code for free Common Core resources for teachers and students, including vetted activities, multimedia, and booklinks, for deeper subject comprehension.

Visit **abdobooklinks.com** or scan this QR code for free additional online weblinks for further learning. These links are routinely monitored and updated to provide the most current information available.

Learn More

Dinosaur Atlas. National Geographic, 2022.

Lippard, Hannah. *Discover Dinosaurs: And What They Could Do*. Flowerpot Press, 2022.

Thomas, Rachael L. *Digging for Dinosaurs*. Abdo, 2019.

Index

brain, 16

extinction, 8

forelegs, 13, 22
fossils, 21–23, 25, 27

Giraffatitan, 25, 26–27
giraffes, 13, 17

Jurassic Park, 24–27

North America, 14, 22

paleontologists, 8, 16, 21, 24, 25
predators, 16

Riggs, Elmer, 21

teeth, 17–18
trees, 7, 13, 16, 17

About the Author

Priscilla An is a writer and editor. She loves talking about dinosaurs with her younger cousins.